Jessica Lizbeth Sanchez Rivas

Stratigraphic profiles of Cs-137 activity in sedimentary rocks

Jessica Lizbeth Sanchez Rivas

Stratigraphic profiles of Cs-137 activity in sedimentary rocks

Nuclei of two Mexican aquatic systems (Lake Santa María del Oro, Nayarit and Laguna de Términos, Campeche)

ScienciaScripts

Imprint

Any brand names and product names mentioned in this book are subject to trademark, brand or patent protection and are trademarks or registered trademarks of their respective holders. The use of brand names, product names, common names, trade names, product descriptions etc. even without a particular marking in this work is in no way to be construed to mean that such names may be regarded as unrestricted in respect of trademark and brand protection legislation and could thus be used by anyone.

Cover image: www.ingimage.com

This book is a translation from the original published under ISBN 978-620-0-01745-1.

Publisher:
Sciencia Scripts
is a trademark of
Dodo Books Indian Ocean Ltd. and OmniScriptum S.R.L publishing group

120 High Road, East Finchley, London, N2 9ED, United Kingdom
Str. Armeneasca 28/1, office 1, Chisinau MD-2012, Republic of Moldova, Europe
Printed at: see last page
ISBN: 978-620-7-40051-5

ABSTRACT

^{137}Cs is an artificial radionuclide released to the environment between 1950 and 1960 due to atmospheric testing of nuclear weapons, its stratigraphic record is used to validate the ^{210}Pb dating, through the maximum ^{137}Cs activities registered in 1963. The main objective of this work was to determine the profiles of ^{137}Cs activity in sediment cores from Lago Santa Maria del Oro, Nayarit (SAMO), and Laguna de Terminos, Campeche (LT), in order to corroborate the ^{210}Pb derived chronology. The two cores from Lago Santa Mana de el Oro (SAMO) were collected with a gravity corer while the two cores from Laguna de Terminos (LTME) were collected manually with PVC tubes. The sediment samples were prepared in calibrated geometry and ^{137}Cs activities were determined by using gamma spectrometry. ^{137}Cs activity interval was higher in SAMO 14-2 core (2.92±1.24 - 23.06±1.65 Bq kg^{-1}) than SAMO 18-4 core (2.29±1.09 - 12.4±1.28 Bq kg^{-1}), and the peak of ^{137}Cs activity showed a better definition in SAMO 14-2. The low ^{137}Cs activities in SAMO 18-4 were attributed to a greater sediment supply in the collection core site, that diluted ^{137}Cs activity. ^{137}Cs activity intervals in LT cores were comparable (LTME1: 1.79±0.52 - 4.16±0.59; LTME2: 2.24±0.36 - 3.99±0.74 Bq kg^{-1}), and ^{137}Cs activity in the stratigraphic profile of LTME2 core do not display a well-defined peak. The low ^{137}Cs activity in the LTME cores is related to the high solubility of ^{137}Cs in seawater. In SAMO 14-2, SAMO 18-4 and LTME1 cores, ^{137}Cs peaks corroborated the ^{210}Pb derived chronology; however, in LTME2 core, the chronology was not validated, due to ^{137}Cs maximum activity does not match with the expected year of 1963, probably related to the supply of eroded soil from the drainage basin. This work demonstrated the useful of ^{137}Cs as stratigraphic marker to corroborate recent dating in sedimentary cores from Mexican aquatic systems.

Keywords:[137] Cs, stratigraphic profile, lacustrine and coastal sediment cores.

INTRODUCTION

Sediment cores are considered environmental archives because they can provide information about past conditions and constitute reliable records of ecosystem evolution. Environmental reconstruction through sediments is based on the assumption that sediment layers are in the same order in which they were deposited (superposition principle), so that, if these layers are not mixed, they reflect the temporal sequence of previous events and it is possible to reconstruct the environmental conditions when they were formed [1].

By studying sediment cores in aquatic environments, information can be obtained on the temporal variation of sediment accumulation rates [2], nutrient fluxes and contaminants such as heavy metals and persistent organic compounds [3].

One of the most widely used radioisotopes for dating recent sediments (~100 years) is ^{210}Pb, however, the chronology derived with this method needs to be corroborated by other independent markers, among which is the radioactive isotope ^{137}Cs, whose sources and processes of transfer to the sediment are independent of those of ^{210}Pb [4].

^{137}Cs, whose half-life ($t_{1/2}$, time for half of the initial amount of the nuclides to decay) is 30.05±0.08 years [5], is an artificial radionuclide that was released into the environment during atmospheric nuclear weapons testing in the late 1950s and early 1960s [6; 7]. The maximum value of ^{137}Cs activity in the sedimentary record can be used to identify the depth corresponding to the period of maximum atmospheric nuclear testing (19631964) [8].

In a stratigraphic profile of ^{137}Cs, a base level or background region is observed corresponding to the period before 1950 (beginning of the nuclear atmospheric tests), from this year the concentrations of^{137}Cs in the sediments increase, a maximum value is presented in ~1963 and later there is a decrease in the concentrations of this radionuclide due to the prohibition of atmospheric tests of thermonuclear weapons [9]. Sometimes, particularly in sediments from Europe, a second substantial record of atmospheric precipitation of^{137}Cs is found due to the Chernobyl nuclear reactor explosion in 1986, and where this occurs, the sedimentary record may also contain a second maximum [10]. Figure 1 shows the stratigraphic profile of^{137}Cs from a sediment core from Lake Espejo de los Lirios (State of Mexico) showing the two peaks in^{137}Cs activity corresponding to 1963 and 1986 [11].

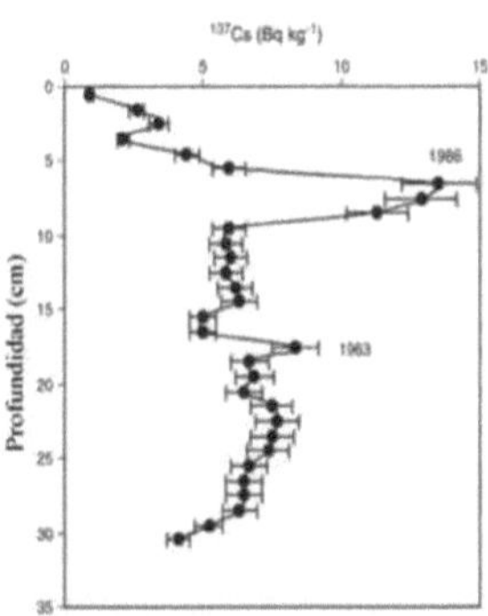

Activity profile of^{137}Cs in Lake Espejo de los Lirios [11].

The main objective of this work was to determine the activity of^{137}Cs in sedimentary cores collected in the Santa Mana del Oro Lake and in marshes associated with the Terminos Lagoon, in order to corroborate the ages determined with the^{210}Pb dating method. Lake Santa Mana del Oro and Terminos lagoon constitute important tunnelsic areas, and several anthropogenic activities such as fishing, trade and mining in Santa Mana del Oro [12] as well as agriculture, livestock and hydrocarbon exploitation in Terminos lagoon [13] are developed in their surroundings.

CONTEXTUAL FRAMEWORK

2.1.   Company

2.1.1.   Historical background

The Instituto de Ciencias del Mar y Limnolog^a of the Universidad Nacional Autonoma de Mexico (UNAM), is a pioneer institution in coastal and marine research in northwestern Mexico. During the late 60's and early 70's, as a result of work carried out by the Department of Marine Sciences and Limnology of the Institute of Biology of the UNAM in a contract with the Secretariat of Hydraulic Resources, the first facilities of the UNAM began to operate, the first facilities of the Mazatlan Station, which were located in Punta Tiburon, and were shared with the School of Marine Sciences of the Autonomous University of Sinaloa (UAS) and with the Regional Center for Fisheries Research (CRIP- SEMARNAP), began to operate.

Later, during the construction period of the first phase of the current facilities, the station facilities were moved to a building on Venustiano Carranza Street. The facilities of the Mazatlan Academic Unit were inaugurated on November 3, 1976. On March 11, 1999, the Consejo Tecnico de la Investigacion CienUfica approved the Internal Regulations of the Instituto de Ciencias del Mar y Limnolog^a. On December 9, 1999, and according to the Diario Oficial de la Federacion, Tomo DLV No. 7, additionally, a 73,547.87 m^a land area was donated to the Instituto de Ciencias del Mar y Limnolog^a[2] located in the Unas estuary, Belvedere Island [14].

2.1.2.   Geographic space

The Mazatlan Academic Unit houses 15 laboratories, a computer center, a library and an aquarium room. The laboratory in which this thesis is developed is the Laboratory of Isotopic Geochemistry and Geochronology (LGIG), was created in 2001 and has the technical and instrumental capacity for the analysis of: (1) natural and artificial radionuclides by high resolution and low background alpha and gamma spectrometry, (2) elemental composition by X-ray fluorescence spectrometry (XRF), (3) C/H/N/S elemental composition, and (4) grain size distribution by laser diffraction.

2.1.3.   Institute and LGIG staff

The academic staff of the Institute of Marine Sciences and Limnology of UNAM is composed of 67 researchers and 53 academic technicians, of which 36 researchers and 26 technicians are assigned to the Ciudad Universitaria campus; 17 researchers and 12 technicians at the Mazatlan Academic Unit in the state of Sinaloa; 12 researchers and 11 technicians at Puerto Morelos in Quintana Roo; and 2 researchers and 3 technicians at the El Carmen station in Ciudad del Carmen, Campeche. The staff of the Geoqwmica Isotopic and Geochronology laboratory consists of Dr. Ana Carolina Ruiz Fernandez who is in charge of the laboratory, M. en C. Libia Hascibe Perez Bernal, laboratory technician [15] and Dr. Tomasa del Carmen Cuellar Marthez, internal advisor of the present thesis.

2.2.   Problem statement

^{210}Pb is a natural radionuclide widely used for dating sedimentary cores; however, the age models obtained need to be validated with independent time markers. ^{137}Cs is an artificial radionuclide frequently used as a stratigraphic marker, which is expected to be detectable in the sediments of Lake Santa Mana del Oro and Laguna de Terminos, and its stratigraphic profiles can be used to corroborate the age of these sediments.

2.3.   Justification

Retrospective analysis of recent environmental changes, based on the analysis of sediment cores, requires the determination of a reliable time frame. Having stratigraphic profiles of ^{137}Cs in the sedimentary cores allows validation of the age models established for the sediments, which confers certainty to the historical reconstructions derived from the analysis of these environmental records.

2.4.   Objectives

2.4.1.   General Objective

To determine the activity of ^{137}Cs, by gamma spectrometry, in samples of sedimentary cores, collected in the Santa Mana del Oro lake, in Nayarit; and in the Terminos lagoon, in Campeche, with the purpose of corroborating the chronology derived with the method of ^{210}Pb of both ecosystems.

2.4.2.   Specific objectives

- To prepare sediment samples in calibrated geometry for the analysis of the specific activity of Cs.^{137}
- Determine the activity of ^{137}Cs in sediments by gamma spectrometry.
- Build the stratigraphic profile of ^{137}Cs activities in the sedimentary cores.
- Verify that the ^{137}Cs profile is useful for corroboration of the chronology derived with the ^{210}Pb method.

THEORETICAL FRAMEWORK

3.1. Radioactivity and radioactive isotopes

Radioactive isotopes (or radionuclides) have unstable nuclei and lose excess energy through the process of radioactive decay, which involves the spontaneous emission of alpha (a) or beta (в) particles and gamma (Y) rays. In this process, the original atom (parent radionuclide) is transformed into the atom of another element that may be stable or radioactive (daughter radionuclide), thus establishing a chain of successive disintegrations (radioactive series) that ends only when a stable element is produced [1].

The types of emissions that occur in the decay process of a radioisotope are as follows:

1. a-particles. A form of radioactive emission identical to the nucleus of helium-4 atoms having atomic number 2 and mass number 4. Thus, with the emission of an alpha-particle the atomic number of the element decreases by 2 units and the mass number by 4 units [16].

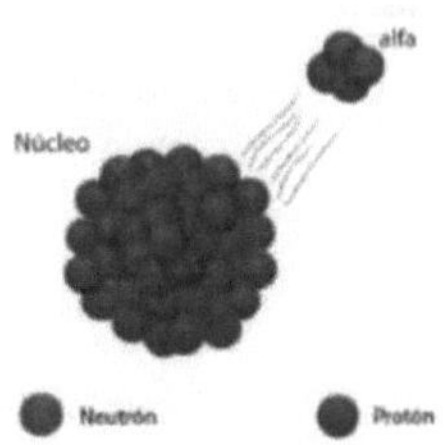

Figura 2. mission of a partícula a.

2. Particles в. A form of radioactive emission that is identical to an electron, it has

a charge of -1. With the emission of a beta particle the atomic number of the element increases by 1 and the mass number remains the same [16].

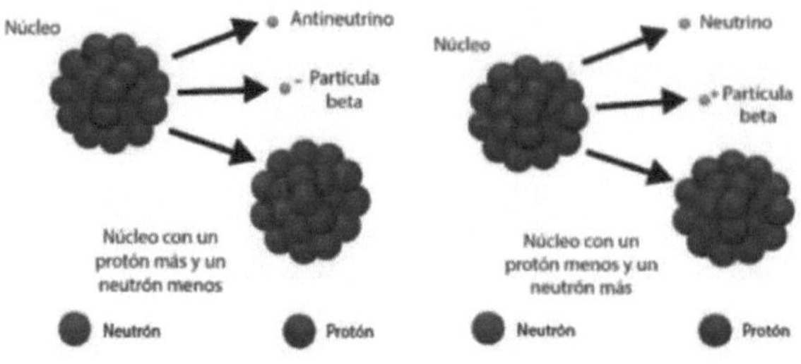

Figura 3. Emission of a particle в- (a) У в+ (b).

3.1. Rays AND. The gamma ray or photon has no charge or mass and therefore has no effect on the atomic number or mass number of an element [16].

3.2. Dating of sediments with^{210} Pb and its corroboration with Cs^{137}

The most commonly used technique to establish a chronological framework in recent (~100 years) lake and coastal sediments is dating with^{210} Pb and corroboration with the artificial radioisotope^{137} Cs [17] which is present in the environment as a result of atmospheric nuclear weapons testing in the years 1954-1964 [10], which falls within the period datable with Pb.^{210}

The radiochronology with^{210} Pb needs to be corroborated by using stratigraphic markers such as the artificial radionuclides^{137} Cs and^{239} Pu which, being products of the atmospheric detonation of nuclear weapons, the maximum of their activities in the sedimentary profiles should coincide with the time horizon from 1962 to 1964, which was the period of greatest global nuclear testing activity [18]. If there is correspondence of the maxima of [137] Cs and [239] Pu with the dates derived from the analysis of [210] Pb, it is proved that the other parameters calculated through radiochronology are reliable [19].

Once the geochronological framework of a sedimentary core is established, it is feasible to reconstruct various aspects of global change, for example, the increase of heavy metal pollution in the environment or persistent organic pollutants (POPs), as well as variations in fluxes and preservation of organic carbon in sediments [1; 20].

3.3.   Factors influencing the deposition of Cs^{137}

The deposition of [137] Cs on the surface of the planet is related to environmental factors such as latitude, rainfall and altitude. Radioactive waste released to the atmosphere from nuclear atmospheric tests was deposited mainly in the northern hemisphere, where most of the nuclear tests were conducted. The largest deposition of [137] Cs was recorded between latitudes 30°- 50° N (Figure 4) [21].

[137]Cs can reach soil or aquatic systems by two different transport mechanisms: dry and wet deposition. The former occurs when pollutants in gaseous phase or attached to particles are deposited on the surface of soil, water or biota, while wet deposition occurs when pollutants are dissolved or suspended in hydrometeor droplets (e.g. rain, snow, fog) that precipitate and impact on the land surface [22]. Atmospheric deposition of [137] Cs is more intense at high altitudes than at lower altitudes because the radioactive elements are deposited directly from the air [23].

The variation in the concentration of [137] Cs in sediments may be related to the time elapsed since its release to the environment, since, due to radioactive decay, the activity of [137] Cs in the sediment has decreased by about 70% of the initially deposited activity [24]. Likewise, the high solubility of [137] Cs in seawater delays its incorporation into the sediment after its arrival in the aquatic environment due to atmospheric deposition [25]. Additionally, [137] Cs can be mobilized into the sediment column through interstitial water as a result of diagenetic processes due to physicochemical processes (e.g. changes in redox conditions) [26]. The concentration of [137] Cs can also be influenced by the presence in the sediments of clay minerals (e.g. Ilites and biotite) that are able to adsorb [137] Cs [27] by cation exchange, i.e., Cs can replace other cations. [137]Cs can replace other cations that are located on the surface or in the interlaminar space of the clay structure, so the higher the content of clay minerals in the sediments, the higher the concentration of 137Cs [28].

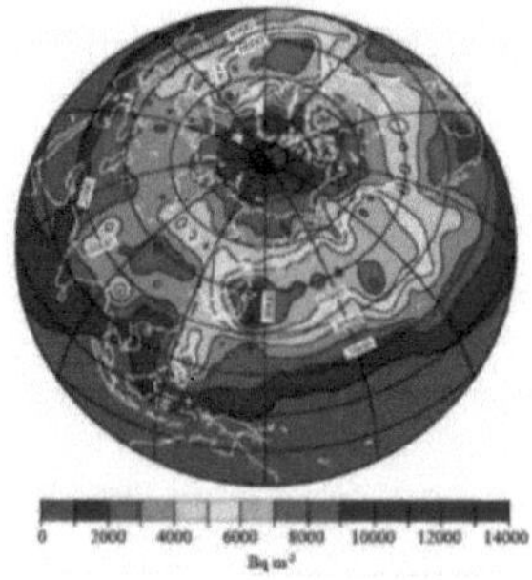

Figura 4. Worldwide geographic distribution of 137Cs [21].

In Mexico, environmental studies have been developed in which [137] Cs maxima have been useful to corroborate age models in lake sediments from Lake Chapala, Jalisco [29] and Lake Espejo de los Lirios, State of Mexico [11], as well as in coastal sediments from Laguna de Terminos, Campeche [30] and Laguna de Alvarado, Veracruz [24]. However, in other studies no maximum of [137] Cs has been detected, but its presence was used to corroborate that the sediments had accumulated in the period after the atmospheric nuclear tests (~1950) [31; 32].

In sediments from Laguna de Mitla, Guerrero [33], not Culiacan, Sinaloa [34], Altata-Ensenada del Pabellon, Sinaloa [35], Gulf of Tehuantepec, Mexico [36; 37], Laguna Ohuira, Sinaloa [38], the activity of [137] Cs was found below background levels, which was associated with the low atmospheric flux of ^{137}Cs in the region because the major deposition of ^{137}Cs occurred at the nearby sites of the atmospheric detonations of nuclear weapons (30-50°N) [39]. In addition, the climatic conditions of low rainfall on the northern Pacific coast of Mexico and the high solubility of ^{137}Cs in seawater are related to the low activities of ^{137}Cs in sediments [39; 34].

STUDY AREA

4.1.   Santa Mana del Oro Lake

Lake Santa Mana del Oro (SAMO) is located in the south of the State of Nayarit (21°22'58" N, 104°34'48" W, Fig. 5a, b) at 750 masl. The lake has an extension of 4 km^2 with a diameter of 2 km and a maximum depth of 65 m; it is endorheic and is fed by precipitation, surface runoff and subsurface flow [40]. It is classified as a warm monomphctic lake (the mixing phase occurs between February and March), it remains stratified most of the year with a thermocline between 17 and 20 m [41]. The lake has mesotrophic characteristics, with high concentrations of phosphorus and sHice [42]. The climate in the lake area is semi-humid sub-humid and temperate sub-humid with summer rainfall, the total annual precipitation is 1,000-1,500 mm and the annual temperature range is 16-26°C [43]. Lake SAMO is surrounded by rural settlements of low to high marginalization, and an urban zone of low marginalization. 23% of the population of the municipality is economically active, of which 63% is dedicated to the agricultural sector and 21% to the service sector. The lake is the main tourist attraction of the municipality, the gradual development as a tourist destination (for at least 40 years) has implied the growth of urbanized areas that for several decades have discharged waste directly into the lake, as well as the proliferation of fossil fuel based boats [44].

4.2.   Terminos Lagoon

Laguna de Terminos (LT, 18° 36' 28.3" N, 91° 33' 21" W) is located in the southern Gulf of Mexico, in the state of Campeche (Fig. 5a, c). It is a coastal lagoon with a surface area of 1,800 km^2 and an average depth of 4 m, which receives water discharges from the Candelaria, Chumpan and Palizada rivers. LT is separated from the open ocean by Isla del Carmen, a barrier with two openings that allow the exchange of saltwater and freshwater, and is home to the city of Carmen (~170,000 inhabitants in 2010) [45], which has developed around the oil industry. The climate of the region is warm sub-humid with abundant summer rainfall, total annual precipitation of 1,200-2,000 mm and mean annual temperature of 26-28°C [43]. Carmen Island is surrounded by approximately 4,392 hectares of mangroves, which include *Rhizophora mangle*, *Avicennia germinans*, *Laguncularia racemosa* and *Conocarpus erectus* [46]. Mangroves provide important environmental services, including coastal protection, water purification, fishery sustenance, and biodiversity conservation. In addition, they have a great capacity to capture and store carbon dioxide (CO_2) in the form of organic carbon, at an annual rate of two to four times greater than that of mangrove forests.

that of tropical forests [47].

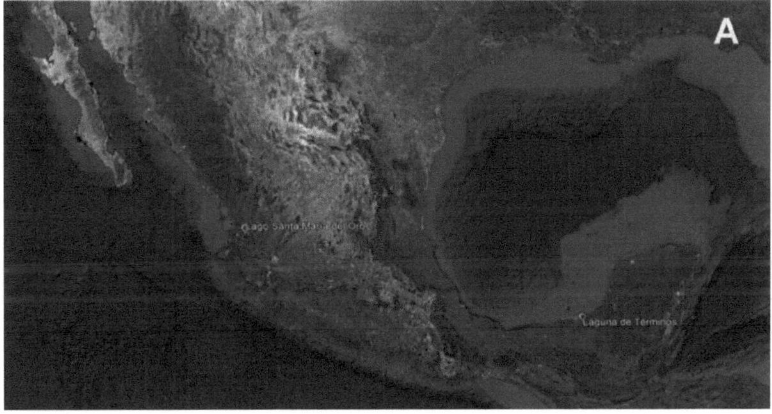

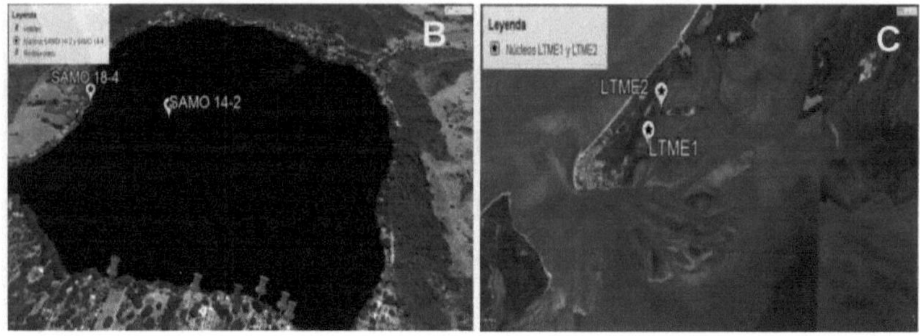

Location of study areas (a) and collection sites of sediment cores SAMO 14-2 and SAMO 18-4 in Lake Santa Mana del Oro, Nayarit (b), and LTME1 and LTME2 in Laguna de Terminos, Campeche (c).

MATERIALS AND METHODS

5.1.   Sampling

5.1.1.   Santa Maria del Oro Lake

In Lake Santa Mana del Oro, two sediment cores (SAMO 14-2 and SAMO 18-4, Figure 5b) were collected with a UWITEC gravity nucleator™ and transparent PVC tubes, 8.6 cm internal diameter and 1.2 meters long. SAMO core 14-2, 78 cm long, was collected at coordinates 21°22'11.20.20"N and 104°34'20.10"W, on April 28, 2014, at a water column depth of 48.2 meters. SAMO core 18-4, 36 cm long, was collected at coordinates 21°22'15.06"N and 104°34'37.30"W, on May 1, 2018, at a water column depth of 6 meters.

5.1.2.   Terminos Lagoon

Two sediment cores were collected manually, with PVC tubes of 10 cm inner diameter and 1 meter in length, at a water column depth of <1 meter, on April 9, 2018 in marshy areas, in the surroundings of the Terminos lagoon (LT). Core LTME1, 59 cm long, was collected at coordinates 18°47'34.84.84"N and 91°27'49.62'W; and core LTME2, 64 cm long, was collected at coordinates 18°48'21.69"N and 91°27'27.30"W (Figure 5c).

5.2.   Sample preparation

The sediment cores were extruded and sectioned at each centimeter. The wet weight of the samples was recorded and they were dried by lyophilization for 72 hours, in a Labconco equipment™ (model no. 7754042), at 0.1 Torrs of pressure and a temperature of approximately -40°C. The dry weight was recorded to obtain the percentage of moisture. One aKquota of each sample was ground in a porcelain mortar and the samples were placed in plastic bags until analysis.

5.3.   Laboratory analysis

For the analysis of ^{137}Cs, an aKquota of powdered sediment (~4 g) was weighed into a 56 mm long, 11 mm diameter polyethylene vial. The vial with the sample was placed in an Ortec™ well HPGe gamma detector for counting for a minimum of 48 hours until an uncertainty of less than 10% was obtained [11]. The detailed procedure for the analysis of ^{137}Cs in sediments is presented in Annex 1.

RESULTS

6.1.   Activities of[137] Cs in sediments

6.1.1.   Lake Santa Mana del Oro

In the SAMO 14-2 core, the activity range of[137] Cs was from 2.92±1.24 to 23.06±1.65 Bq kg[-1] . The maximum activity value was recorded in the 16-17 cm section (23.06±1.65 Bq kg[-1] , Figure 6a).

In the SAMO 18-4 core, the activity range of[137] Cs was from 2.29±1.09 to 12.4±1.28 Bq kg[-1] . The maximum activity value was recorded in the 29-30 cm section (12.4±1.28 Bq kg[-1] , Figure 6b).

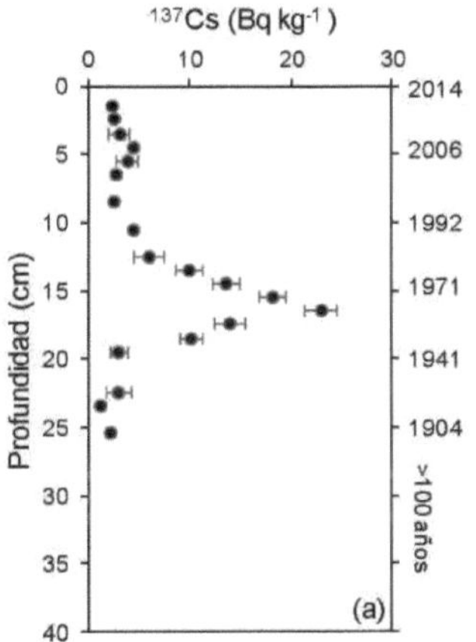

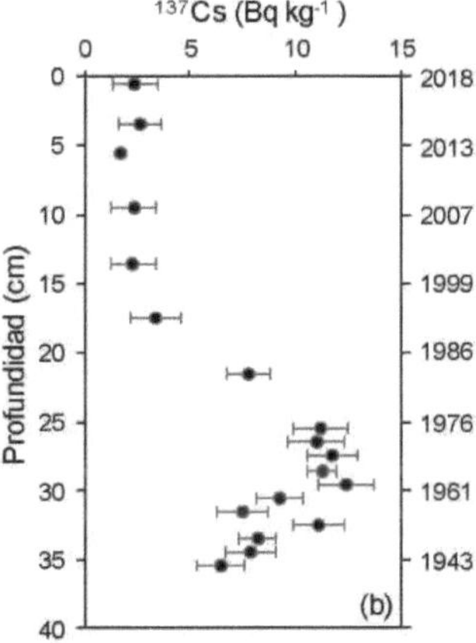

According to the age model (CF) with[210] Pb obtained for both nuclei [48], the sections in which the highest activity of[137] Cs was detected corresponded to the years 1957±6.3 - 1964±6.1 for the SAMO 14-2 core and 1965±3.37 - 1968±3.08 for the SAMO 18-4 core, which coincide with the period of maximum nuclear atmospheric testing (1962 - 1964) [31], thus validating the dating with Pb.[210]

Figure 6. Stratigraphic profiles of[137] Cs activity in the nuclei: SAMO 14-2 (a) and SAMO 18-4 (b).

6.1.2.   Terminos Lagoon

In the LTME1 core, the activity range of[137] Cs was from 1.79±0.52 to 4.16±0.59 Bq kg[-1] . The maximum activity value was observed in the 12-13 cm section (4.16±0.59 Bq kg[-1] , Figure 7a) which, according to the age model obtained with[210] Pb, corresponded to the years 1962±3.1 - 1964±2.8 [48] which coincides with the maximum of atmospheric nuclear testing (1962 - 1964) [37] and therefore validates the chronology obtained with the[210] Pb method.

In the LTME2 core the activity range of[137] Cs was from 2.24±0.36 to 3.99±0.74 Bq kg[-1] . The maximum activity value was found in the 19-20 cm section (3.99±0.74 Bq kg[-1] , Figure 7b), which according to the CF dating model corresponded to the years (1930±5.2 - 1934±4.6) [48] which does not coincide with the maximum of atmospheric nuclear tests and therefore does not validate the dating model with Pb.[210]

Figure 7. Stratigraphic profiles of[137] Cs activity of the nuclei: LTME1 (a) and LTME2 (b).

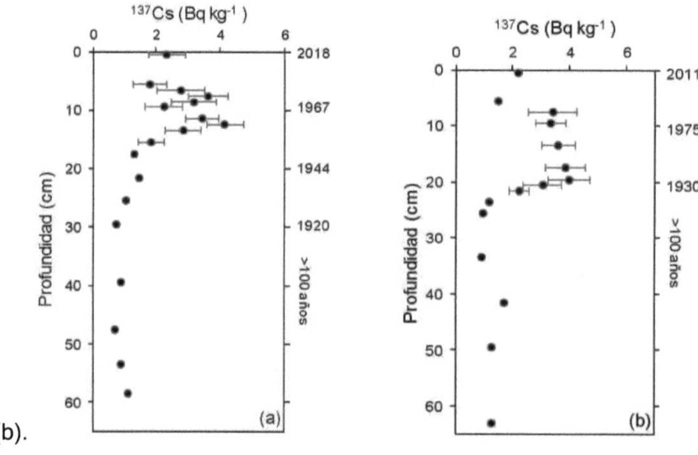

DISCUSSION OF RESULTS

7.1.   Activities of Cs^{137}

The activity of^{137} Cs in SAMO 14-2 (2.92±1.24 - 23.06±1.65 Bq kg^{-1}), was higher than that recorded for the SAMO 18-4 core (2.29±1.09 to 12.4±1.28 Bq kg^{-1}). The lower activity of^{137} Cs in SAMO core 18-4, may be related to a higher sediment supply, since this core was collected in an area closer to the SAMO lake shoreline, and the mass accumulation rate was seven times higher (0.15±0.04 to 0.88±0.24 g cm^{-2} anno^{-1}) than in SAMO core 14-2 (0.011±0.003 to 0.77±0.004 g cm^{-2} anno^{-1}) [48]. In addition, the activity profile of^{137} Cs in the sediments of SAMO core 14-2 showed better definition than in SAMO core 18-4. The maximum values of both cores were found in sections 16-17 cm in SAMO 14-2 and 29-30 cm in SAMO 18-4, which, according to the age model, correspond to the years 1957 - 1964 and 1965 - 1968 respectively, in agreement with the expected maximum in 1963.

The activity of^{137} Cs in the nuclei of Lake SAMO was similar to that recorded for Lake Espejo de los Lirios, Estado de Mexico (3-13 Bq kg^{-1}) [11] and Lake Zirahuen, Michoacan (8-18 Bq kg^{-1}) [31] and was lower than that determined for Lake Chapala, Jalisco (30-50 Bq kg^{-1}) [29] and Lago Verde, Veracruz (10-107 Bq kg^{-1}) [49] (Table 1).

The lower activity of^{137} Cs in Lake SAMO with respect to Lake Chapala may be related to differences in altitude, which is lower in Lake SAMO (1173 masl) than in Lake Chapala (1548 masl). Atmospheric deposition of^{137} Cs at high altitudes is higher because it is deposited directly from the air [23]. In the case of the lower activity of^{137} Cs in Lake SAMO than in Lake Verde, it may be due to differences in rainfall, which is twice as high in Lake Verde (2,401 mm anno^{-1}) than in Lake SAMO (1,132 mm anno^{-1}), and the deposition of^{137} Cs is higher in sites with high levels of rainfall [50].

In the sedimentary cores of Laguna de Terminos, the activity of [137] Cs was comparable (LTME1: 1.79±0.52 to 4.16±0.59 Bq kg^{-1} ; LTME2: 2.24±0.36 to 3.99±0.74 Bq kg-1). The maximum values of 137Cs were found in sections 12-13 cm in LTME1 and 19-20 cm in LTME2, which, according to the age model, correspond to the years 1962 - 1964 and 1930 - 1934 respectively, so that only the chronology derived from the [210] Pb method in the LTME1 core was corroborated; the absence of a well-defined maximum of [137] Cs in the stratigraphic profile of the LTME2 core may be related to the deposition of eroded soils from the drainage basin.

The activity of [137] Cs in both cores was similar to that recorded in sediments from Laguna de Alvarado, Veracruz (1.20-4.60 Bq kg^{-1} [24]) and was within the range of activities determined in some coastal environments in Mexico, such as the Coatzacoalcos estuary, Veracruz (1.80-7.70 Bq kg [32]; Gulf of Tehuantepec, Oaxaca (1-1.70 Bq kg [36]), as well as in sediments from the Philippine Sea (1.20-7.70 Bq kg [36]).80-7.70 Bq kg^{-1} [32]); Gulf of Tehuantepec, Oaxaca (1-1.70 Bq kg^{-1} [36]), as well as in sediments from the Philippine Sea (0.20-2.80 Bq kg^{-1} [51]), Sumba Island, Indonesia (<1.50 Bq kg^{-1} [52]) and Southeast Atlantic Ocean (0.30-1.79 Bq kg^{-1} [53]) (Table 1). However, it was lower than that obtained in sediments from the Black Sea, Turkey (37.45 Bq kg^{-1} [54]), due to the deposition of [137] Cs associated with the Chernobyl accident at the latter site [54].

The low activity of [137] Cs in coastal sediments is related to the high solubility of this radionuclide in seawater, which has been observed in sediment cores from the northern part of Mexico (Culiacan estuary, the Altata-Ensenada del Pabellon lagoon and the Ohuira lagoon in Sinaloa) [34; 35; 38].

Table 1. Activities of [137] Cs in sediments of different aquatic systems in Mexico and around the world.

Type of environm	Site	Activity of	Coordinates of	Rainfall[a] (mm year	Altitude[d]

ent		^{137}Cs (Bq kg)$^{-1}$	the core	)$^{-1}$	(masl)	Reference
Lakes	Lake Chapala, Jalisco	30-50	20°150'N; 103°000'O	859±227	1548	[29]
	Lake Zirahuen, Michoacan	8-18	19°26'N; 101°44'O	1,069±368[b]	2075	[31]
	Lake Espejo de los Lirios, Mexico City	6-13.50	19°38'N; 99°13'O	664±186	2280	[11]
	Lago Verde, Veracruz	10-107	18°36.72'N; 95°20.87'O	2,401±557	149	[49]
	Lake Santa Mana del Oro, Nayarit	2.92-23.06 2.29-12.40	21°22'11.2"N; 104°34'20.1"O 21°22'15.0"N; 104°34'37.3"O	1,132±354	750	Present study
Lakes	Xingyung Lake	1-8.90	24°20'19"N; 102°46'51"E	n.a.	1722	[55]
	Lake Puyehue, Chile	2.50-11.80	40°40'S; 72°24'W	n.a.	211	[57]

Type of environment	Site	Activity of ^{137}Cs (Bq kg)^{1}
Coastal lagoon	Mitla Lagoon, Guerrero	-8.33
		21-40
	Laguna de Terminos, Campeche	1.79-4.16
		2.24-3.99
	Altata-Ensenada del Pabellon, Sinaloa	0.02
	Ohuira Lagoon, Sinaloa	Level background

	Alvarado Lagoon, Veracruz	1.20-4.60
	River Estuary	
Estuaries	Culiacan, Sinaloa	0.33

Core coordinates	Rainfall[3] (mm year)[1]	Altitude[d] (masl)	Reference
n. d	1,331±415	0	[33]
18°40'N; 91°30'O			[30]
18°47'34.8 "N; 91°27'49.6 "O	1,244±331	0	Present study
18°48'21.7 "N; 91°27'27.3 "O			
n. d	412±236	0	[35]
25°41'N; 108°53'O	323±149	1	[38]
18°47'52.4 "N; 95°51'28.4 "O	1,703±455	1	[24]
n. d	169±213[c]	1	[34]

Type of environment	Site	Activity of 137Cs (Bq kg1)
	River Estuary	
Estuaries	Coatzacoalcos, Oaxaca	1.80-7.70
		1.40-3.90

Rios	Rio Aar y Rhine River, Switzerland	10-50
Sea inside	Yanhe River, China	0.92-4.82
	Black Sea, Turkey	2.08-37.45
Platform continental	Go lfo de Tehuantepec, Mexico	Background level 1-1.70
	Equatorial Pacific Western, Philippine Sea	1.30-2.80 0.20-0.50

Core coordinates	Pluvial precipitation3 (mm year1)	Altitude (masl)	Reference
n. d			[32]
	1,708±734	90	
18°11.103'N; 94°27.078'0			
n. d	n.a.	1877	[56]
n. d	n.a.	1050	[58]
n. d	n.a.	45	[54]
15°27.22'N; 94°22.86'O 15°59.987'N; 94°48.469'0 7°51.50'N;	2,184±745	0	[36] [37]
126°33.28'E 8°0.03'N; 126°34.50'E	n.a.	0	[52]

Type of environment	Site	Activity of 137Cs (Bq kg-1)	Coordinates of the core	Rainfall precipitation (mm anno-1)	Altitude (masl)	Reference
Platform continental	Sumba Island, Indonesia	<1.50	n.a.	n.a.	0	[52]
	Southeast Ocean Atlantico, Brazil.	0.30-1.79	28°40'N	n.a.	0	[53]

n. d: not available.

a total annual average for the period ~1960-2016, 1947-2016b and 1981 -2008c obtained from the nearest meteorological station [59].

d altitude obtained from getamap.net [60].

7.2. Relationship of latitude, rainfall and altitude to deposition of Cs^{137}

7.2.1. Latitude

The activity of[137] Cs in sediments varies with latitude, because most of the atmospheric nuclear weapons tests (between 1950s and 1960s) were conducted at sites in the northern hemisphere, so that radioactive debris was dispersed from these sites to other latitudes through atmospheric circulation [6]. Globally, the largest deposition of[137] Cs has been recorded at latitude 45° in the northern hemisphere (Table 2), with an average value of 5,090 Bq m^{-2} [21]. Because Mexico is located between 14° 30' N and 32° 43' N of the northern hemisphere, much of the country is outside the latitudes where the largest atmospheric deposition of[137] Cs occurred [21]. We compared the activity of[137] Cs in sediments vs. latitude of different aquatic systems (Figure 8) and did not find a significant correlation (R=0.11, p>0.05), probably due to other variables such as altitude, rainfall, salinity of the medium and clay content in the sediment, which can influence the deposition and accumulation of[137] Cs, as well as diagenetic processes that can promote the diffusion of[137] Cs through the sediment column [61]. Table 2. Global deposition of[137] Cs according to latitude [21].

Latitude (°)	Average (Bq m-2)	Range (Bq m-2)
Northern		
45	5090	1540-10230
55	5040	2770-7910
35	4100	700-10630
65	3420	1110-8060
25	2620	120-6430
Southern		
-5	800	100-1210
-35	580	150-1430

Gulf of Tehuantepec

Mitla Lagoon. Guerrero

° Coatzacoalcos River Estuary. Oaxaca

Terminos Lagoon, Campeche

Lago Verde. Veracruz

Alvarado Lagoon. Veracruz

Lake Zirahuen. Veracruz

Lake Espejo de Lirios. Mexico City

Lake Chapala, Jalisco

Altata Ensenada del Pabellon. Sinaloa

Culiacan River Estuary, Sinaloa

Ohuira Lagoon

Terminos Lagoon. Campeche

Lake Santa Maria del Oro, Nayarit

Xingyung Lake. China

Lake Puyehue, Chile

Western Equatorial Pacific. Turkish Sea

Southeast Atlantic Ocean. Brazil

Figure 8. Correlation of activities with respect to latitude in aquatic systems of Mexico and the world.

7.2.2. Rainfall precipitation

In this study, the activity of [137] Cs vs. rainfall of different Mexican aquatic systems was evaluated (Figure 9), which was obtained from the total annual rainfall values recorded in the Climatological Stations Network of the National Water Commission (CONAGUA) of the National Meteorological Service (SMN). A significant correlation was observed between the activity of [137] Cs and rainfall ($P<0.05$, $R=0.6214$), this coincides with that recorded in the basins of the Carrion and Noguera Pallaresa rivers in Spain, since a significant correlation was determined between the inventories of [137] Cs and the average annual rainfall [62]. In the oceanic region of the northern hemisphere, the highest deposition of [137] Cs was found in areas with high stratosphere-troposphere exchange and high rainfall rates, while low deposition of [137] Cs is observed in arid and desert regions [21]. Furthermore, Simon et al. (2004) estimated that the deposition of [137] Cs in the United States is higher in the eastern and mid-western region, where rainfall is higher than in the southwestern region, where it is higher than in the western

region, where it is higher than in the western region [20].

arid conditions predominate [63]. In sedimentary cores collected in coastal lagoons

of northwestern Mexico (Estuario del Rfo Culiacan [34], Altata Ensenada del

Pabellon [35], Laguna de Ohuira [38] in Sinaloa), the activity of [137] Cs did not exceed

the baseline level and in that region rainfall is scarce (Table 1). Although the absence

of the [137] Cs signal in coastal sediments is also related to the high solubility of the

radionuclide in seawater [39].

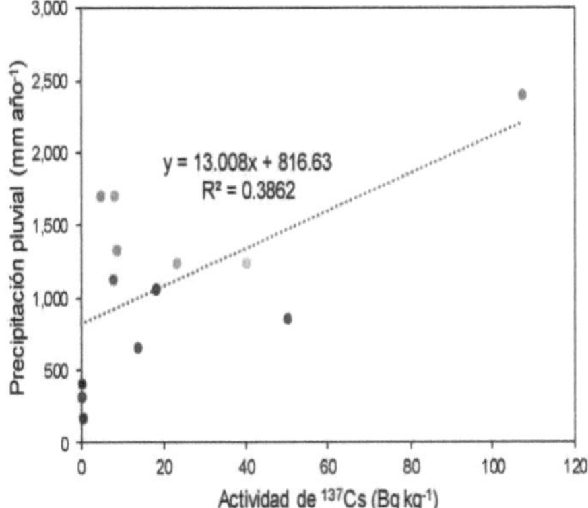

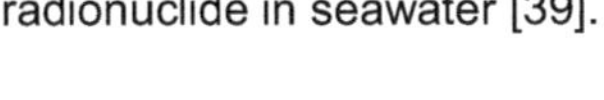

Figure 9. Correlation of activities with respect to precipitation in aquatic systems of

7.1.1.   Altitude

The activity of [137] Cs in sediments was compared against altitude in different aquatic

systems in Mexico (Table 1) and a significant relationship was found (P<0.05,

R=0.6291) (Figure 10), which is due to the fact that atmospheric precipitation is more

intense at high altitudes, because contaminants are deposited directly from the air

[23]. This was also observed by Kubica et al. (2002) who analyzed soil samples in the Tatra National Park, Poland and the highest activities of[137] Cs were determined at sites located at an altitude >1300 masl [64].

Figure 10. Activity of[137] Cs with respect to altitude in aquatic systems in Mexico.

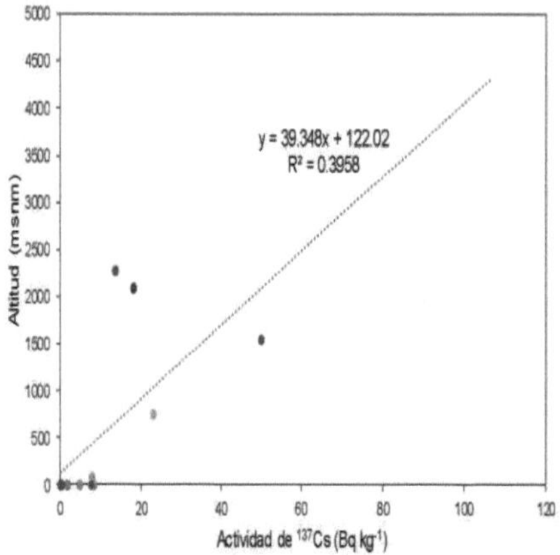

CONCLUSIONS

1. The preparation of sediment samples and analysis of the radionuclide activity [137] Cs in the SAMO 14-2 and SAMO 18-4 cores collected in Lake Santa Mana del Oro and the LTME1 and LTME2 cores from Laguna de Terminos were carried out.

2. The range of activities of [137] Cs was determined in the four cores analyzed, which was higher in the sediment cores of Lake Santa Mana del Oro (2.29±1.09 to 23.06±1.65 Bq kg^{-1}) than in Laguna de Terminos (1.79±0.52 - 4.16±0.59 Bq kg)$.^{-1}$

3. The stratigraphic profile of [137] Cs was constructed for each of the analyzed cores. The maximum activity of [137] Cs in cores SAMO 14-2, SAMO 184 and LTME1, was useful to validate the chronologies derived with the [210] Pb dating method, while for core LTME2, the chronology could not be validated, since the maximum activity of [137] Cs did not coincide with the expected date of 1963, probably due to the supply of eroded soils from the drainage basin.

4. Based on data of [137] Cs activities in sediments of different aquatic systems in Mexico published in the literature, and the one obtained for the cores analyzed in this research work, a significant positive correlation was determined between the activities of [137] Cs with rainfall and the altitude of the sampling zones, since the atmospheric deposition of [137] Cs in the sediments occurs mainly through the humid v^a and in the higher zones the radionuclide is deposited directly from the air.

BIBLIOGRAPHY

[1] A.C Ruiz-Fernandez, J. Sanchez-Cabeza, F. Paez-Osuna, J. Ontiveros-Cuadras, "Environmental records of global change," Science and Development, 2014.

[2] P. Lopez, "Recent geochronology with^{210} Pb of the accumulation of organic carbon and persistent organic compounds (PAHs and PCBs) at two sites on the Mexican Pacific Continental Shelf," Master's Thesis, Universidad Nacional Autonoma de Mexico, Mazatlan, Sinaloa, Mexico, 2011.

[3] A.C. Ruiz-Fernandez, "Distribucion espacial y temporal de metales pesados en sedimentos lacustres de la cuenca Mexico: Chalco, Texcoco y Cuautitlan Izcalli, Estado de Mexico," Doctoral Thesis, Universidad Nacional Autonoma de Mexico, Mazatlan, Sinaloa, Mexico, 204, 1999.

[4] J. Sanchez-Cabeza, M. D^az-Ascencio, A. Ruiz-Fernandez, "Radiochronology of coastal sediments using^{210} Pb: Models, validation and applications," 2012.

[5] Marie-Martine Be, V. Chiste, C. Dulieu, *Table of Radionuclides* (3), 91, 2006. Available at: www.nucleide.org. Accessed 05/09/2018.

[6] J. Bernal, L. Beramendi, K. Lugo-Ibarra, L. Daessle, "Revision to some radiometric geochronometers applicable to the Quaternary," Boleth de la Sociedad Geologica Mexicana. V. 3 (62), pp. 305-323, 2010.

[7] S. Krishnaswami, D. Lal, *Radionuclide limnochronology*, In A. Lerman (ed.), Lakes: Chemistry, Geology, Physics. Springer-Verlag, New York, 1978, pp. 153177.

[8] Q. HE, D. E. Walling, "Interpreting Particle Size Effects in the Adsorption of[137] Cs and Unsupported[210] Pb by Mineral Soils and Sediments," Environmental Radioactivity 30, pp. 117-137, 1996.

[9] R. D. Delaune, W. H. Patrick, R. J. Buresh, "Sedimentation rates determined by[137] Cs dating in a rapidly accreting salt marsh," United States, Lousiana: Nature. 275, pp. 532-533, 1978.

[10] S. F. Sugai, M. J. Alperin, W. S. Reeburgh, "Episodic deposition and[137] Cs immobility in Skan Bay sediments: a ten-year[210] Pb and[137] Cs time series," Marine Geology 116, pp. 351-372, 1994.

[11] A. C. Ruiz-Fernandez, F. Paez-Osuna, J. Urrutia-Fucugauchi, M. Preda, "[210] Pb geochronology of sediment accumulation rates in Mexico City Metropolitan Zone as recorded at Espejo de los Lirios lake sediments," Catena 61 (1), pp. 31-48, 2005.

[12] mexicoo.mx, "Actividades economicas y servicios en Santa Mana Del Oro," in *Directorio empresarial mexicano,* 2018. [On line]. Available at: https://mexicoo.mx/empresas/nayarit/santa-maria-del-oro/actividades-economicas [Accessed: January 9, 2019].

[13] Ecured, "Laguna de Terminos," in *Media Wiki*, 2015. [Online]. Available from: https://www.ecured.cu/Laguna_de_Terminos [Accessed: January 9, 2019].

[14] UNAM, Instituto de Ciencias del Mar y Limnolog^a, "Antecedentes," 2018. Available at: http://www.icmyl.unam.mx/mazatlan/uves/mazatlan/es/quienes-we are/background

[15] UNAM, Instituto de Ciencias del Mar y Limnolog^a, "Personal Academico," 2018. Available at: http://www.icmyl.unam.mx/mazatlan/uves/mazatlan/es/personal-academic-mztn

[16] W. Daub and W. S. Seese, "Qumica Nuclear" in *Qumica*, Pearson Eucacion, Mexico, 2005, pp. 652-653.

[17] E.D. Goldberg, M. Koide, "Geochronological studies of deep-sea sediments by the ionium/thorium method: Geochimica et Cosmochimica," Acta, 26, pp. 417-450, 1962.

[18] S. Krishnaswamy, D. Lal, J.M. Martin, M. Meybek, "Geochronology of lake sediments: Earth and Planetary Science Letters," 11: pp. 407-414, 1971.

[19] A.C. Ruiz-Fernandez, F. Paez-Osuna, M.L. Machain-Castillo, E. Arellano-Torres, "^{210}Pb geochronology and trace metal fluxes (Cd, Cu and Pb) in the Gulf of Tehuantepec, South Pacific of Mexico," Journal of Environmental Radioactivity. 76 (1-2), pp. 161-175, 2004.

[20] Ontiveros-Cuadras, J. "Retrospective study of the historical trends of fluxes of potentially toxic elements (As, Cr, Cu, Hg, Pb, Rb, Zn, V), persistent organic pollutants (PCBs, PBDEs and PAHs) and organic carbon, in two lakes located in the Mexican Altiplano, characterized by contrasting levels of anthropization," Universidad Nacional Autonoma de Mexico, 2015.

[21] M. Aoyama, K. Hirose, Y. Igarashi, "Re-construction and updating our understanding on the global weapons tests^{137} Cs fallout," J. Environ. Monit., 8, pp. 431-438, 2006.

[22] J. E. Figueruelo and M. Marino-Davila, "Transferencia en compartimientos medioambientales" in *Qumica ftsica del ambiente y de los procesos medioambientales,* Reverte, Espana, 2014, pp. 257.

[24] A. C. Ruiz-Fernandez, M. Maanan, J. Sanchez-Cabeza, L. Perez-Bernal, P. Lopez, A. Limoges, "Chronology^a of recent sedimentation and geochemical characterization of the sediments of the Alvarado lagoon, Veracruz (southwestern Gulf of Mexico)," Marine Sciences, 40 (4), pp. 291-303, 2014.

[25] M. Casas-Ruiz, M. Barrera, F. Feria, C. Corredor and R. A. Ligero, "Un Modelo de Migracion Vertical del[137] Cs," Geoqumica Isotopica Aplicada al Medioambiente, Seminarios de la Sociedad Espanola de Mineralog^a, 1, pp. 107-147, 2004.

[26] P. A. Lima Ferreira, E. Siegle, C. A. Franca Schettini, M. Michaelovitch, R. C. Lopes Figueira, "Statistical validation of the model of diffusion-convection (MDC) of[137] Cs for the assessment of recent sedimentation rates in coastal systems," J Radioanal Nucl Chem, pp. 2059-2071, 2015.

[27] H. Mukai, A. Hirose, S. Motai, R. Kikuchi, K. Tanoi, T. M. Nakanishi, T. Yaita, and T. Kogure, "Cesium adsorption/desorption behavior of clay minerals considering actual contamination conditions in Fukushima," Scientific Reports, pp. 1-7, 2016.

[28] M. Okumuraa, S. Kerisitb, I. C. Bourgc, L. N. Lammersde, T. Ikedaf, M. Sassib, K. M. Rossob, M. Machidaa, "Radiocesium interaction with clay minerals: Theory and simulation advances Post-Fukushima," Journal of Environmental Radioactivity 189, pp. 135-145, 2018.

[29] F. Fernex, P. Zarate-del Valle, H. Rammez-Sanchez, F. Michaud, C. Parron, J.

Dalmasso, G. Barci-Funel, M. Guzman-Arroyo, "Sedimentation rates in Lake Chapala (Western Mexico): possible active tectonic control," Chemical Geology, 177, pp. 213-228, 2001.

[30] J. C. Lynch, J.R. Meriwether, B.A. McKee, F. Vera-Herrera, R.R. Twilley, "Recent accretion in mangrove ecosystems based on ^{137}Cs and ^{210}Pb," Estuaries. 12 (4), pp. 284-299, 1989.

[31] S. J. Davies, S.E. Metcalfe, A.B. MacKenzie, A.J. Newton, G.H. Endfield, J.G. Farmer, "Environmental changes in the Zirahuen basin, Michoacan, Mexico, during the last 1000 years," Journal of Paleontology. 31 (1), pp. 77-98, 2004.

[32] A. C. Ruiz-Fernandez, J. Sanchez-Cabeza, A. Hernandez, V. Martmez-Herrera, H. Perez-Bernal, M. Preda, Hillaire-Marcel, J. Gastaud, A. Quejido-Cabezas, "Effects of land use change and sediment mobilization on coastal contamination (Coatzacoalcos River, Mexico)," Continental Shelf Research, 37, pp. 57-65, 2012.

[33] F. Paez-Osuna and E.F. Mandelli, "^{210}Pb in a tropical coastal lagoon sediment core," Estuarine, Coastal and Shelf Science, 20, pp. 374-387, 1985.

[34] A.C. Ruiz-Fernandez, C. Hillaire-Marcel, B. Ghaleb, F. Paez-Osuna, M. Soto Jimenez, "Recent sedimentary history of anthropogenic impacts on the Culiacan River Estuary, northwestern Mexico: geochemical evidence from organic matter and nutrient," Environmental Pollution, 118, pp. 365-377, 2002.

[35] A.C. Ruiz-Fernandez, C. Hillaire-Marcel, B. Ghaleb, F. Paez-Osuna, M. Soto Jimenez, "Isotopic constraints (^{210}Pb, ^{228}Th) on the sedimentary dynamics of contaminated sediments from a subtropical coastal lagoon (NW Mexico),"

Environmental Geology, 41, pp. 74-89, 2001.

[36] A.C. Ruiz-Fernandez, F. Paez-Osuna, M.L. Machain-Castillo, E. Arellano-Torres, "^{210}Pb geochronology and trace metal fluxes (Cd, Cu and Pb) in the Gulf of Tehuantepec, South Pacific of Mexico," Journal of Environmental Radioactivity. 76 (1-2), pp. 161-175, 2004.

[37] A. C. Ruiz-Fernandez, C. Hillaire-Marcel, A. Vernal, M. Machain-Castillo, L. Vasquez, B. Ghaleb, Aspiazu-Fabian, F. Paez-Osuna, "Changes of coastal sedimentation in the Gulf of Tehuantepec, South Pacific Mexico, over the last 100 years from short-lived radionuclide measurements," Estuarine, Coastal and Shelf Science 82, pp. 525-536, 2009.

[38] A. C. Ruiz-Fernandez, M. Frignani, T. Tesi, H. Bojorquez-Leyva, L. Bellucci, F. Paez Osuna, "Recent sedimentary history of organic matter and nutrient accumulation in the Ohuira Lagoon, Northwestern Mexico," Archives of Environmental Contamination and Toxicology, 53 (2), pp. 159-167, 2007.

[39] A. C. Ruiz-Fernandez, C. Hillaire-Marcel, "^{210}Pb-derived ages for the reconstruction of terrestrial contaminant history into the Mexican Pacific coast: potential and limitations," Mar. Pollut. Bull. 59 (4), pp. 134-145, 2009.

[40] S. Sosa-Najera, S. Lozano-Garrta, P.D. Roy, M. Caballero, "Record of historical droughts in western Mexico based on elemental analysis of lake sediments: The case of Lake Santa Mana del Oro," Boletm de la Sociedad Geologica Mexicana 62, 3, pp. 437-451, 2011.

[41] D. Serrano, A. Filonov, I. Tereshchenko, "Dynamic response to valley breeze

circulation in Santa Mana del Oro, a volcanic lake in Mexico," Geophysical Research Letters, 29, 13, pp. 1649, 2002.

[42] M. Caballero, A. Rodnguez, G. Vilaclara, B. Ortega, Roy Priyadarsi, S. Lozano, "Hydrochemistry, ostracods and diatoms in a deep, tropical, crater lake in Western Mexico," 72 (3), pp. 512-523, 2013.

[43] CONABIO, "National Commission for the Knowledge and Use of Biodiversity," 2018. Available at: http://www.conabio.gob.mx/conocimiento/regionalizacion. Accessed 11/09/2018.

[44] SEDATU, "Atlas de Riesgos del Municipio de Santa Mana del Oro, Nayarit," Secretary of Agrarian, Territorial and Urban Development. Government of Nayarit, 372.

[45] INEGI, "XIII General Population and Housing Census 2010," 2010. Available at: http://www.inegi.gob.mx/. Accessed 12/09/18.

[46] H.G. Reyes-Gomez and A.D. Vazquez-Lule, "Caracterizacion del sitio de manglar Isla del Carmen, in Comision Nacional para el Conocimiento y Uso de la Biodiversidad (CONABIO)," Sitios de manglar con relevancia biologica y con necesidades de rehabilitacion ecologica, CONABIO, Mexico, D.F., 19, 2009.

[47] CEC, "Blue Carbon in North America," Commission for Environmental Cooperation, Montreal, pp. 4, 2014.

[48] J. M. Garrta, "Geochronology with ^{210}Pb in sedimentary cores from two aquatic systems in Mexico. Marismas de la Laguna de Terminos, Campeche and Lago Santa

Mana del Oro, Nayarit," Licenciatura Thesis, Universidad Politecnica de Sinaloa, Mazatlan, Sinaloa, Mexico, 2019.

[49] A. C. Ruiz-Fernandez, C. Hillaire-Marcel, F. Paez-Osuna, B. Ghaleb, and M. Caballero, "^{210}Pb chronology and trace metal geochemistry at Los Tuxtlas, Mexico, as evidenced by a sedimentary record from the Lago Verde crater lake," Quaternary Research 67, pp. 181-192, 2007.

[50] R. Martmez Lugo, G. Felipe Oliva, "La contaminacion radioactiva de los ecosistemas," 1997.

[51] D. Pittauer, P. Roos, J. Qiao, W. Geibert, M. Elvert, H.W. Fischer, "Pacific Proving Ground radioisotope imprint in the Philippine Sea sediments," Journal of Environmental Radioactivity 186, pp. 131-141, 2018.

[52] D. Pittauer, S.G. Tims, M.B. Froehlich, L.K. Fifield, A. Wallner, S.D. McNeil, H.W. Fischer, "Continuous transport of Pacific-derived anthropogenic radionuclides towards the Indian Ocean," scientific reports, pp. 1-8, 2017.

[53] C.L Figueira, M.G. Tessler, M.M de Mahiques, I.L. Cunha, "Distribution of ^{137}Cs, ^{238}Pu and $^{239+240Pu}$ in sediments of the southeastern Brazilian shelf-SW Atlantic margin," Science of the Total Environment 357, pp. 146-159, 2006.

[54] Hasan Baltasa, Murat Sirina, Goktug Dalgicb, Ugur Cevikc, "An overview of the ecological half-life of the ^{137}Cs radioisotope and a determination of radioactivity levels in sediment samples after Chernobyl in the Eastern Black Sea, Turkey," Journal of Marine Systems, 177, pp. 21-27, 2018.

[55] C. Gao, J. Yu, X. Min, A. Cheng, R. Hong, and L. Zhang, "Heavy metal concentrations in sediments from Xingyun lake, southwestern China: implications for environmental changes and human activities," Environmental Earth Sciences, pp. 1-13, 2018.

[56] J. Abrahama, K. Meusburgerb, J. Kobler Waldisb, M.E. Kettererc, M. Zehringera, "Fate of[137] Cs,[90] Sr and[239+240] Pu in soil profiles at a water recharge site in Basel, Switzerland," Journal of Environmental Radioactivity, 182, pp. 85-94, 2018.

[57] F. Arnaud, O. Magand, E. Chapron, S. Bertrand, X. Boës, F. Charlet, M.A. Melieres, "Radionuclide dating (^{210} Pb,^{137} Cs,^{241} Am) of recent lake sediments in a highly active geodynamic setting (Lakes Puyehue and Icalma-Chilean Lake District)," Science of the Total Environment 366, pp. 837-850, 2006. [58] [59] [60]

[59] SMN, Servicio Meteorologico Nacional, "Normales Climatologicas por Estado," 2018. Available at: http://smn.cna.gob.mx/es/climatologia/informacion-climatologic Accessed: 25/09/2018

[60] "Maps of the whole world (Latitude and Longitude)," 2018. Available at: http://es.getamap.net/ Accessed: 27/09/2018.

[61] P. Porto, Des E. Walling, G. Callegari, and A. Capra, "Using caesium-137 and unsupported lead-210 measurements to explore the relationship between sediment mobilization, sediment delivery and sediment yield for a Calabrian catchment,"

58 X. Zhang, D.E. Walling, Q. Yang, X. He, Z. Wen, Y. Qi, M. Fen, "[137] Cs budget during the period of 1960s in a small drainage basin on the Loess Plateau of China,"
Journal of Enviromental Radioactivy 86, pp. 78-91, 2006.

Marine and Freshwater Research 60, pp. 680-689, 2009.

[62] J. A. Sanchez-Cabeza, M. Garrta-Talavera, E. Costa, V. Pena, J. Garrta-Orellana, P. Masque, C. Nalda, "Regional calibration of erosion radiotracers (^{210}Pb and ^{137}Cs): atmospheric fluxes to soils (northern Spain)," Environmental Science and Technology 41 (4), pp. 1324-1330, 2007.

[63] S. L. Simon, A. Bouville and H. L. Beck, "The geographic distribution of radionuclide deposition across the continental US from atmospheric nuclear testing," J. Environ. Radioact, 74, pp. 91-105, 2004. [61]

ANNEXES

Annex 1. Procedure for the analysis of[137] Cs by gamma spectrometry.

1. Materials and equipment

a) Materials:

PVC hook-and-loop fastener

BoKgraph

Plastic spoons

Compactor (5 ml syringe embolus)

Rubber stoppers

Teflon tape

4 ml capacity polyethylene tubes

Adhesive Tape

Laboratory logbook

b) Equipment

Drying stove

Analytical balance

Ortec HPGe Gamma Spectrometer HPGe

Kquido Nitrogen Recirculator (Mobius)

2. Procedure for sample preparation

    2.1.   **Drying the samples**

Sediments to be analyzed should be pre-dried in an oven at 45 °C for ~24 hours.

    2.2.   **Cooling the samples**

The dried samples are removed from the oven and placed in a desiccator with sHica gel; wait at least 1 hour for the sediments to cool and reach room temperature.

2.3.   **Tube filling**

To determine the weight of the sample to be analyzed, use an analytical balance and proceed as follows:

- Mark polyethylene tubes to desired volume (2 or 4 ml) depending on sample availability.
- Tare the polyethylene pipes.
- Fill the tubes with the sediment up to the reference line (approximately 2 or 4 g, depending on the geometry e.g. 2 mL or 4 mL).
- Compact the sediment with the plastic plunger.
- Weigh tubes filled with sample
- Record the weight in the logbook.
- Close the tubes with a rubber stopper and seal with Teflon tape.
- Label the tubes with adhesive tape, specify the name and section of the nucleus, as well as the weight of the sediment.
- Coat the tubes with hook-and-loop fasteners.
- Allow 21 days to elapse to allow radioactive equilibrium between ^{222}Rn and Pb.^{214}

3. Measurement procedure by gamma spectrometry

3.1.   **Start-up:**

- Open the GammaVision software and in the **File** menu, select **Settings and** in the **Sample description** option specify the detector on which the sample will be analyzed, the geometry (2 or 4 ml), the nucleus and the sample section (example: G1 4ml SILVA E6B 5-6); in the **Sample size** option specify the mass and unit of measurement of the sample (kg or l). Select **OK**
- Choose **GO** and verify the name and mass of the sample.

3.2.   **Recommendation on measurement time:**

- Measure 2 days.

- Analyze the spectrum (see instructions in "Analysis" below).

- For[137] Cs or[241] Am: under "Summary of Radionuclides" see the column headed Det_Limit (detection limit) and Det_Thres (cntic limit-threshold value). If the activity of the sample is below the cntic Kmite, it is not necessary to count more time; if it is in between, measure one more day; if it is above, stop the measurement.

4. Measurement results analysis procedure

4.1.   On the hard disk of the computer associated with the gamma equipment and in the

the **GAMMA** folder subfolder **Spectra**, check for the existence of a folder of the project and of the core to be analyzed, otherwise create such folders.

4.2.   In the Spectra folder open the "Model Spreadsheet" file.

Gamma Spectrometry **Sediment Profiles 16052016.xls**" in the **File** menu choose the **Save As** option select the folder of the core to be analyzed and name the file as Gamma Spectrometry Results, specify the name of the core e.g. "**Gamma Spectrometry Results LTME1**".

4.3.   In the **File** menu, select **Save as and** automatically opens the **Spectra** folder, find the project folder and corresponding kernel. Set the sections as xx-yy, check the sample name and weight, choose **OK**.

4.4.   In the **Analyze** menu, choose **Settings**, select **Sample type** and in the **File - Browse**: select the appropriate **Sample type** file (e.g. G1 4ml 18052015 sediments marshes.Sdf); choose **OK**. Check the analysis and calibration file (**Calibrate**, **Recall calibration**, choose calibration according to detector, geometry and sample type) by choosing **Open**, or double-click on the corresponding file. To set the sample weight in the **File** menu and in the **Setting** option enter the sample weight and select **OK**.

4.5. In the **Analyze** menu choose **Entire spectrum in memory**: the output is

a RPT format file with the results of the analysis, check the analysis data: name, detector, calibration file; release; check the mass.

4.6. Select the data from the **Summary of nuclides** and copy them to the sheet

MS Excel result sheet for each core, according to the core section analyzed (see section 4.2). Fill in the data in the first row of the sheet (sample name, depth, measurement date, detector and measurement time).

[23] J.W. Mietelski, B. Kubica, P. Gaca, E. Tomankiewicz, S. Blazej, M. Tujeta-Krysa, M. Stobinski, "$^{23}{}_{8,2}2^{39+40}$ Pu,^{241} Am,^{90} Sr and^{137} Cs in mountain soil samples from the Tatra National Park (Poland)," Journal of Radioanalytical and Nuclear Chemistry 275, 3, pp. 523-533, 2008.

I want morebooks!

Buy your books fast and straightforward online - at one of world's fastest growing online book stores! Environmentally sound due to Print-on-Demand technologies.

Buy your books online at
www.morebooks.shop

Kaufen Sie Ihre Bücher schnell und unkompliziert online – auf einer der am schnellsten wachsenden Buchhandelsplattformen weltweit! Dank Print-On-Demand umwelt- und ressourcenschonend produziert.

Bücher schneller online kaufen
www.morebooks.shop

Printed by Books on Demand GmbH, Norderstedt / Germany